ELECTRICITY

by Clint Twist

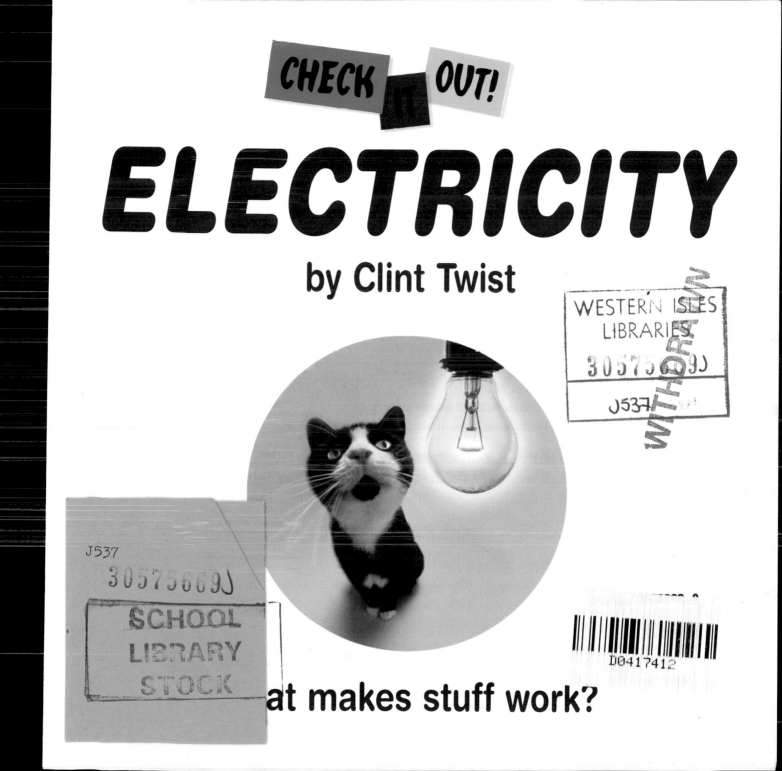

at makes stuff work?

What is electricity?

Electricity is a kind of energy that we use in many different ways.

Electricity is used to make heat and **light**. Many houses have electric heating and lighting.

House lights = electricity

Now it's your turn...

Electricity is used to make lots of different machines work. DVD players, televisions and computers all use electricity.

Torch light

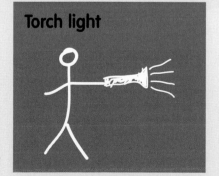

When you turn on a torch it makes light. What kind of energy do you think a torch uses to make light?

Light bulbs

An electrical **light bulb** is a hollow, glass object that produces light.

The small light bulbs used in torches make a small amount of light.

The large light bulbs used in lamps and room lights make much more light.

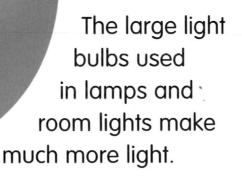

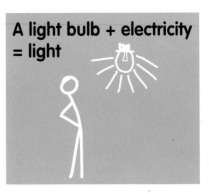

A light bulb + electricity = light

Inside a light bulb is a thin, metal wire.

When electricity goes through the wire it makes the wire very hot. The wire gets so hot that it glows brightly.

A light bulb + no electricity

What happens when there is no electricity going through the wire in the light bulb?

7

Batteries

Batteries are a safe way to store small amounts of electricity. Every battery has two areas of bare metal. These are usually at the ends of the battery.

The metal ends are marked with the symbols "**+**"and "**−**".

A battery has to be put into a torch the right way around, or the torch will not work.

Battery right way around = torch working

Now it's your turn...

The right way is with the + and − on the battery matching the + and − on the torch.

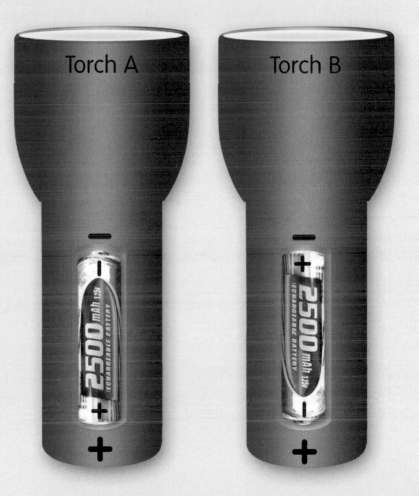

Torch A

Torch B

Torch not working

Do you think both torch A and torch B will work?

Circuits

An electrical **circuit** is a non-stop pathway that electricity can flow around.

Electrical wires are used to connect objects to make a circuit.

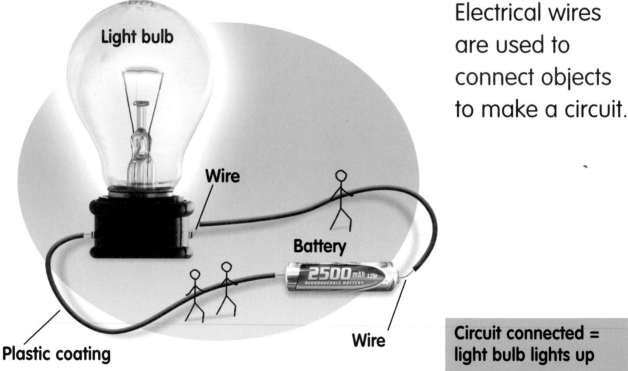

Light bulb

Wire

Battery

2500mAh 1.25V
RECHARGEABLE BATTERY

Wire

Plastic coating

A simple electrical circuit has a light bulb and a battery joined by wires.

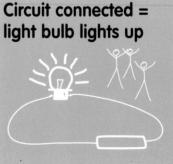

Circuit connected = light bulb lights up

Now it's your turn...

In a circuit, electricity flows from the battery, along the wire, through the light bulb and back to the battery.

Oh dear!

What do you think is wrong with this circuit?

Switches

An electrical **switch** is used to control the flow of electricity around a circuit.

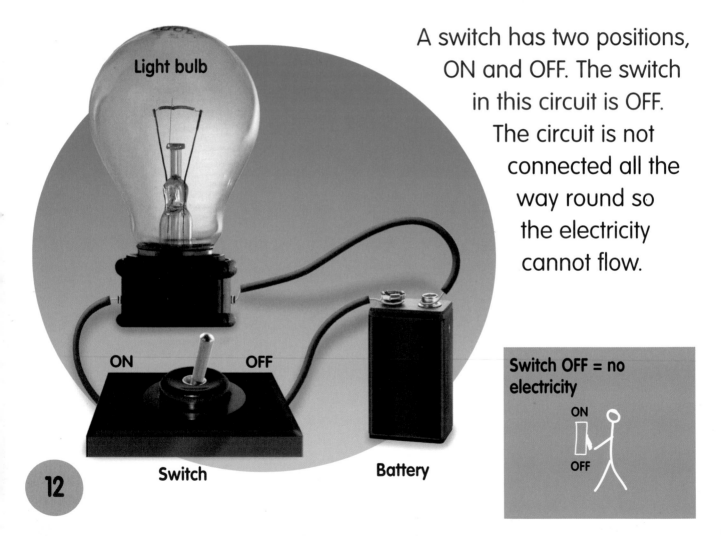

Light bulb

A switch has two positions, ON and OFF. The switch in this circuit is OFF. The circuit is not connected all the way round so the electricity cannot flow.

ON   OFF

Switch OFF = no electricity

ON

OFF

Switch

Battery

Check it out

Now it's your turn...

Metals su...
because e...

Steel wire

Metal wire

Switch ON

Putting the switch in the ON position connects the circuit.

ON   OFF

When the switch is turned OFF, the light bulb is not lit. What happens when the switch is turned ON?

Answers

Electricity

These m

Glas

Page 5

A torch uses electricity to make light.

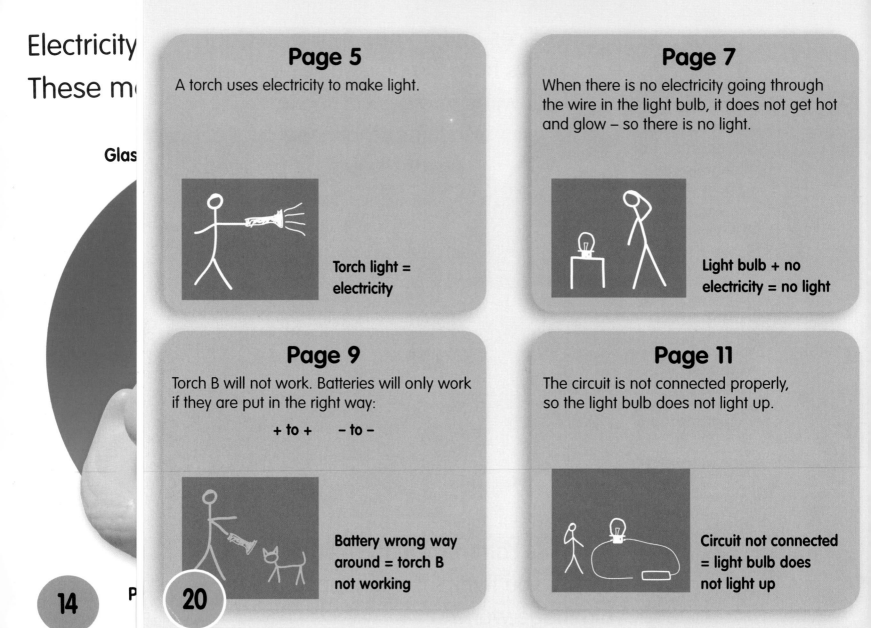

Torch light = electricity

Page 7

When there is no electricity going through the wire in the light bulb, it does not get hot and glow – so there is no light.

Light bulb + no electricity = no light

Page 9

Torch B will not work. Batteries will only work if they are put in the right way:

+ to +   – to –

Battery wrong way around = torch B not working

Page 11

The circuit is not connected properly, so the light bulb does not light up.

Circuit not connected = light bulb does not light up

Page 13

When the switch is ON, it connects the circuit. This allows electricity to flow, lighting the light bulb.

Switch ON = electricity

Page 15

If you use metal wire, the light bulb will light up because metal wire is a good conductor of electricity.

Metal wire = good conductor

Page 17

Metal cannot be used as an insulator because it is a good conductor!

Metal = bad insulator

Page 19

Batteries are not strong enough to make a fridge work. A fridge uses mains electricity from an electrical socket.

Fridge = electricity from a socket

Glossary

batteries Objects that store small amounts of electricity.

circuit The non-stop pathway, or route, that electricity flows along.

conductors Materials that electricity can flow through.

electricity A kind of energy that is used to make heat and light and to power machines.

insulator Any material that electricity cannot flow through.

light A kind of energy that we see with our eyes.

light bulb A hollow glass object containing a thin wire that gives off light when electricity flows through it.

mains electricity The electricity used by homes, offices, factories and shops.

metals Hard, strong, shiny substances, such as steel and copper. Metals are good electrical conductors.

power lines Thick wires that deliver electricity to homes, offices, factories and shops.

pylons Tall, metal towers that are used to hold up power lines.

sockets Small boxes where appliances and machines, such as fridges or computers, can be plugged into the mains electricity supply.

switch An object that is used to connect or disconnect an electrical circuit. Switches are used to turn electrical machines such as computers on and off.

Index

24

Picture credits
t=top, b=bottom, c=centre, l=left, r=right, OFC=outside front cover
Corbis: 6, 7, 17. Photodisc: 18. Powerstock: OFC, 1, 2, 3, 4, 5, 8, 13, 14, 15, 16, 19.

Every effort has been made to trace the copyright holders, and we apologise in advance for any unintentional omissions. We would be pleased to insert the appropriate acknowledgements in any subsequent edition of this publication.